THE LIBRARY OF THE PLANETS™

SATURN

Amy Margaret

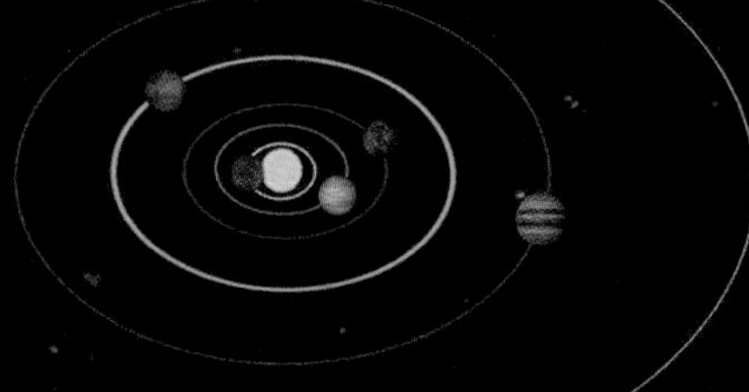

The Rosen Publishing Group's
PowerKids Press™
New York

For Annie Elizabeth and her mom and dad

The author would like to thank Barry Horton for his helpful information on viewing the night sky.

Published in 2001 by The Rosen Publishing Group, Inc.
29 East 21st Street, New York, NY 10010

First Edition

Book Design: Michael Caroleo and Michael de Guzman

Photo Credits: pp. 1, 4, 11, 15, 16 PhotoDisc; p. 7 (Saturn) PhotoDisc, p. 7 (Roman god Saturn) Michael R.Whelan/NGS Image Collection; p. 8 PhotoDisc (digital manipulation by Michael de Guzman); p.10 (*Voyager*) PhotoDisc, p. 10 (Saturn) PhotoDisc; p. 12 PhotoDisc (digital manipulation by Michael de Guzman); p. 19 (Saturn) PhotoDisc, p. 19 (*Huygens* illustration) Courtesy of NASA/JPL/California Institute of Technology; p. 20 (Saturn) Courtesy of NASA/JPL/California Institute of Technology, p. 20 (Hubble Space Telescope) © NASA/Roger Ressmeyer/CORBIS.

Margaret, Amy.
Saturn/ by Amy Margaret.—1st. ed.
p. cm.— (The library of the planets)
Includes index.
Summary: Examines the history, movement, atmosphere, rings, moons, and exploration of the planet Saturn.
ISBN 0-8239-5646-6 (lib. bdg. : alk. paper)
1. Saturn (Planet)—Juvenile literature. [1. Saturn (Planet)] I. Title. II. Series.

QB671 .M37 2001
523.46–dc21 99-088048

Manufactured in the United States of America

Contents

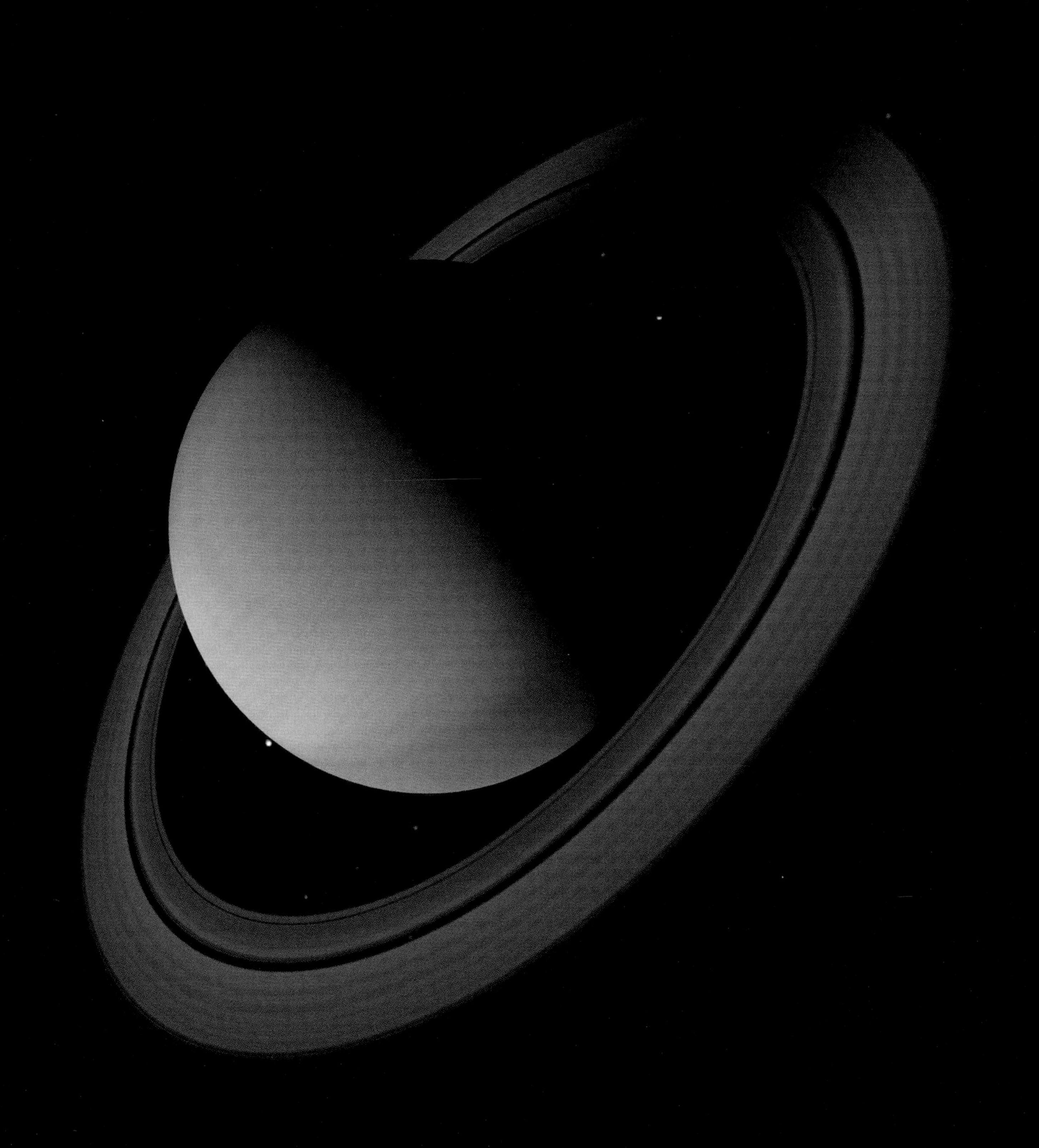

Saturn, the Universe's Beauty

Saturn is the second largest planet in the **solar system**. Only Jupiter is larger. Saturn is so big that if it were hollow, you could fit about 750 Earths inside. Saturn is made mostly of hydrogen and a bit of helium. These are the two most common elements found in the **universe**. The universe is made up of many **galaxies**. A galaxy is a group of stars. Any of these stars may have its own solar system. Our Sun is a star with nine planets in its solar system. Earth and Saturn are two of these planets.

Saturn is the sixth planet from the Sun. It can be found between Jupiter and Uranus. Saturn is best known for the amazing rings that circle the planet. Since these fascinating rings were discovered in the 1600s, **astronomers** have continued to study them.

Saturn is 74,914 miles (120,562 km) across, making it the second biggest planet in the solar system. Only Jupiter is bigger.

The History of Saturn

The planet Saturn was spotted in ancient times, along with Mercury, Venus, Mars, and Jupiter. Before **telescopes** and binoculars were invented, Saturn was thought to be the farthest planet in our solar system. Astronomers could only see Saturn as a tiny spot. The next planet, Uranus, was not seen until 1690. People used to think that Uranus was a star.

Galileo Galilei was the first astronomer to see Saturn through a telescope in 1610. When Galileo looked at Saturn, it seemed like the planet had ears.

Almost 50 years later, in 1656, another astronomer, Christiaan Huygens, saw what Saturn's "ears" really were. With a more advanced telescope, Huygens saw that they were rings around the planet. Saturn was the first planet discovered to have rings around it.

This is an illustration of Saturn and its rings. The planet is named after the Roman god, Saturn, who is the god of the harvest. He is shown here with a farm instrument called a sickle behind him.

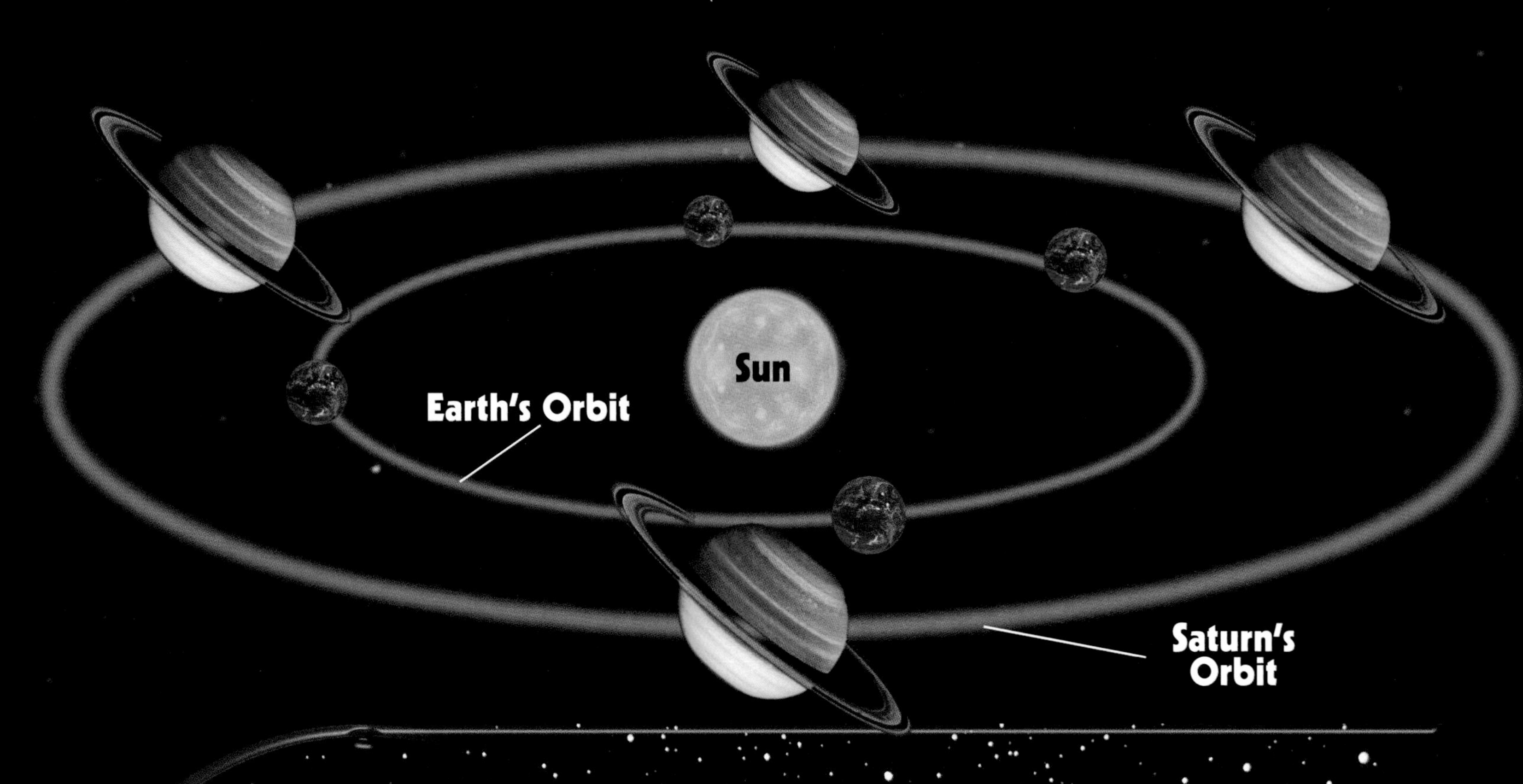

Planet	Orbit Time Around Sun
Mercury	88 Earth Days
Venus	225 Earth Days
Earth	365 Days
Mars	687 Earth Days
Jupiter	12 Earth Years
Saturn	29 Earth Years
Uranus	84 Earth Years
Neptune	165 Earth Years
Pluto	249 Earth Years

Saturn on the Move

All nine planets in our solar system move in two ways. First each planet spins on its **axis**. When a planet spins, it looks like a basketball spinning on the finger of a basketball player. Saturn spins quicker on its axis than all other planets except Jupiter. Saturn makes a complete rotation in about 10 1/2 hours. Earth takes 24 hours, or one Earth day.

The second way the planets move is around the Sun. The Sun pulls the nine planets toward itself, which makes them move. The farther away a planet is from the Sun, the longer it takes for the planet to **orbit** the giant star. Saturn circles the Sun in 29 1/2 Earth years. This is equal to one year on Saturn.

This is a computer image of Saturn orbiting the Sun. The inner ring shows Earth's orbit around the Sun.

Early Journeys to Saturn

In 1979, the **space probe** *Pioneer-Saturn* was the first mission to take close-up pictures of Saturn and its rings. A space probe travels in space and is steered by scientists on the ground. The photos were sent to Earth through *Pioneer*'s computer system. The next two missions were *Voyager 1* and *Voyager 2*. The *Voyager* missions showed that the major rings of Saturn are made up of thousands of thinner rings. They also discovered seven new moons around Saturn. Each *Voyager* space probe measured **atmosphere** and **temperature**. *Voyager 1* also took pictures of Jupiter, and *Voyager 2* photographed Jupiter, Uranus, and Neptune. All three space probes are traveling out of the solar system. The *Voyager* probes are still working! Scientists expect these probes to continue working until around the year 2015.

It took Voyager 1 (above) nearly three years to reach Saturn.

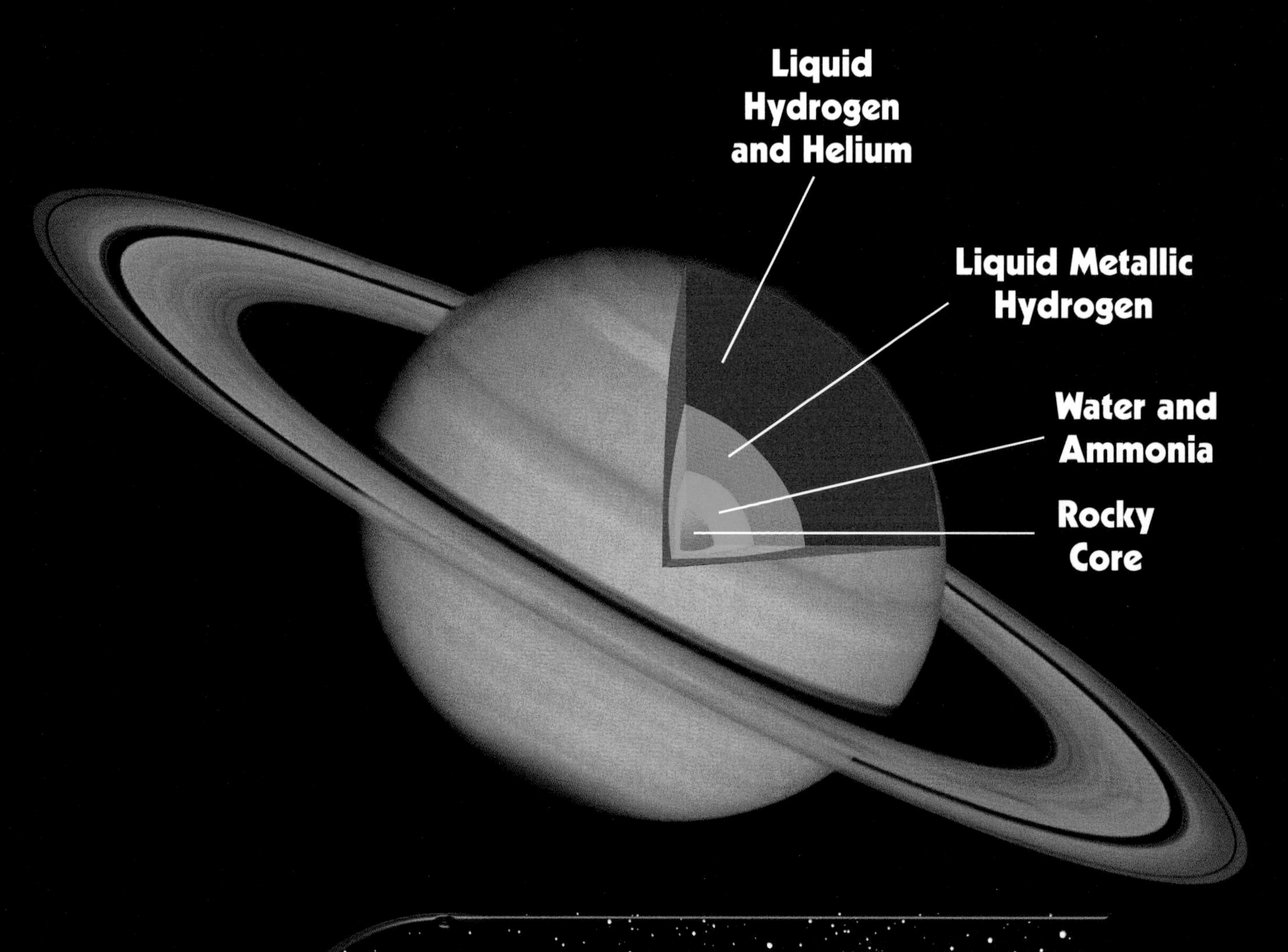

Winds blow continuously on Saturn. Some winds blow very fast. The *Voyagers* measured the winds at about 1,100 miles (1,770 km) per hour in some spots on Saturn.

Under Saturn's Clouds

Saturn, Jupiter, Uranus, and Neptune are sometimes called the gas giants. This is because these planets are made of mostly liquids and gases. This makes Saturn much different from the inner planets, Earth, Mercury, Venus, and Mars. The inner planets are mostly rock and metal. They are called the inner planets because they are the planets closest to the Sun.

Under Saturn's thick atmosphere, scientists think that there are areas where hydrogen and helium gases become liquids. The heavy atmosphere presses down on the hydrogen and helium, turning them into liquids. Water and a gas called ammonia are also present on Saturn. Most scientists believe that Saturn has a rocky **core** at its center. This is different from Earth, which has both liquid and solid iron in its core.

Saturn is so far away from the Sun that it gets very little heat. Temperatures on Saturn are about 300 degrees Fahrenheit (184 degrees C) below zero.

Saturn's Mysterious Rings

Thousands of rings circle the planet Saturn. They make up seven larger rings. The rings are made up of ice pieces and rocks. Some of the pieces are bigger than a house, and others are the size of dust bits. The rings around Saturn were probably formed from the explosion of a moon or **satellite**. A collision of two satellites could also have caused enough rock and ice bits to form rings. In between two large rings is an area called the Cassini Division. This space looks empty, but it isn't. The Cassini Division just has less ice and rocks in it than Saturn's rings. The Cassini Division was named in 1675 after its discoverer, Giovanni Cassini.

Some of Saturn's rings are very wide. They measure as much as 45,000 miles (72,420 km) around, or the distance of about two trips around Earth. The rings seem as thin as paper when looking at them from the side. They actually vary from about 30 feet (9.1 m) to 500 feet (152.4 m) thick.

As viewed from Earth, Saturn appears to have only three rings. It actually has thousands! We call the three rings we see the A-ring, B-ring, and C-ring. The Cassini Division is between the A-ring and the B-ring.

How Many Moons Are In Our Solar System?
Astronomers have been looking for moons in our solar system as long as they have been finding planets. Here is the latest moon count:

- Mercury: 0
- Venus: 0
- Earth: 1
- Mars: 2
- Jupiter: 16
- Saturn: 18
- Neptune: 15
- Uranus: 8
- Pluto: 1

The Many Moons of Saturn

Saturn has at least 18 moons, while Earth has just one. Only nine of Saturn's moons are large enough to be spotted with telescopes. Other moons were discovered in 1980 and 1981 by the *Voyager* space probes. The closest moon to Saturn is Atlas, which is 48,000 miles (77,249 km) away. Phoebe, possibly Saturn's farthest moon, is more than 6 1/2 million miles (10.5 million km) from Saturn! The *Voyager* probes took pictures of some of the moons, including Enceladus, Iapetus, Tethys, and Mimas. Each moon is very different in appearance. The largest moon orbiting Saturn is Titan. Titan is the second largest known moon in our solar system.

This picture of Saturn and some of its moons was made using photographs taken by the Voyager 1 space probe. Although Titan is Saturn's largest moon, it looks small here because it is far away from the planet.

The Giant Titan

Saturn's largest moon is Titan. It is almost 3,200 miles (5,150 km) wide. That's bigger than the entire length of the United States. Titan is part ice and part rock. Titan is different from all other moons we know about today. It has a thick atmosphere. Titan's atmosphere is made up mostly of the gas nitrogen, like Earth, but is thicker. The thick atmosphere keeps scientists from seeing what lies on this moon's surface.

The *Voyager* space probes were not able to photograph through Titan's thick haze. The *Cassini* mission is another space project that will try to find out more about this interesting moon. When *Cassini* reaches Saturn's atmosphere in 2004, it will release a small space probe called *Huygens*. *Huygens* will head toward Titan, taking photos along the way. It is scheduled to land on Titan's surface in 2004 and send back pictures of what it finds.

This is a photo of the Huygens *space probe moving toward Titan. Titan is the second biggest moon in our solar system.*

None of Saturn's other moons are even close to Titan in size. The next largest moon, Iapetus, is only about 950 miles (1,529 km) wide.

If you weigh 100 lbs. (45.4 kg) on Earth, you would weigh 107 lbs. (48.5 kg) on Saturn.

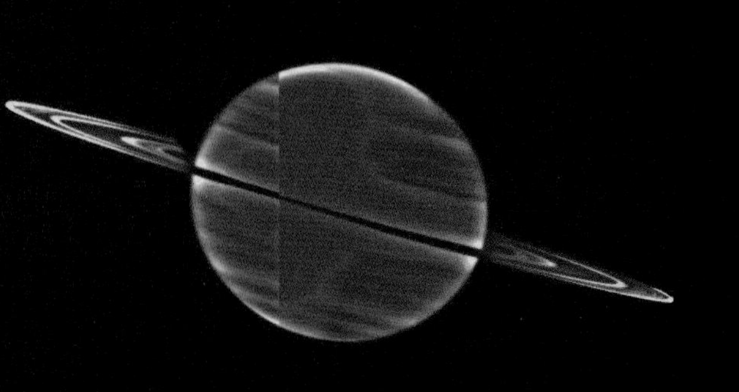

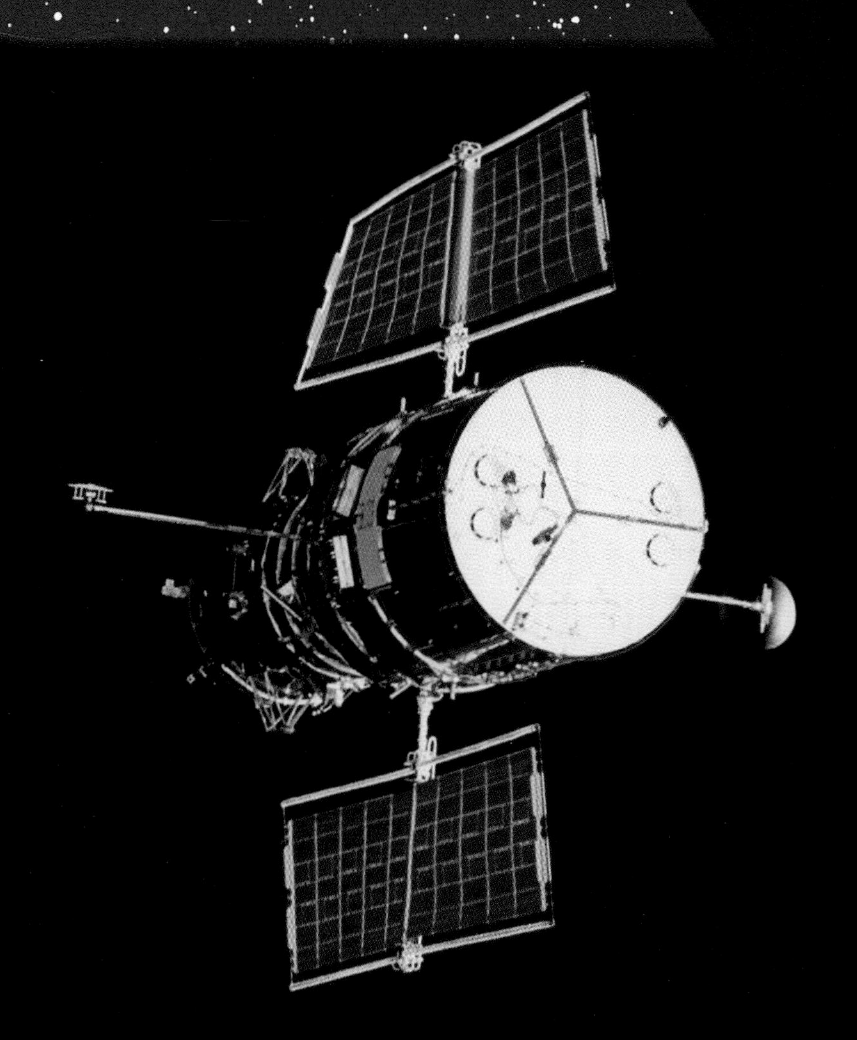

Seeing Saturn From Earth

Saturn is the farthest planet from Earth that can still be seen with the naked eye. This beautiful ringed planet can be viewed from Earth with few or no tools.

With a telescope, you will be able to see a lot more of the planet. You can spot the rings around the planet. You may also be able to track Saturn's largest moon, Titan.

Over the next several years, the best times to view Saturn will be in a dark sky in the winter months. You can also look in an astronomy magazine or on a Web site for more specific times and locations. When you go planet gazing, first find a dark location. Allow your eyes to adjust to the darkness. It might take a while before you start spotting the planets and moons. The more you study the night sky, the better you will become at finding its many wonders.

This picture of Saturn was taken by the Hubble Space Telescope, shown on the bottom. The Hubble Space Telescope orbits Earth to take pictures of objects in our solar system.

The *Cassini* Mission and Beyond

The latest mission to study Saturn is the *Cassini* mission. NASA (National Aeronautics and Space Administration) and ESA (European Space Agency) are working together to carry out this mission. *Cassini* was launched in 1997. It will travel 2.2 billion miles (3.5 billion km) to reach Saturn. The spacecraft is expected to enter Saturn's atmosphere in the year 2004. It will fly around Saturn and its moons for four years.

Scientists hope to learn more about Saturn and its moons. *Cassini* will collect information and photograph Saturn's moons. It will study Saturn's rings and its atmosphere, too. Scientists programmed *Cassini* to release a small space probe to land on Titan. They hope to learn more about Saturn's largest moon, especially its surface.

There is so much more we can learn about Saturn, its moons, and its beautiful rings. Hopefully, the *Cassini* mission will be just one of many future space missions to Saturn, the universe's beauty.

Glossary

astronomers (ah-STRAH-nuh-merz) People who study the night sky, the planets, moons, stars, and other objects found there.
atmosphere (AT-muh-sfeer) The layer of gases that surrounds an object in space. On Earth, this layer is the air.
axis (AK-sis) A straight line on which an object turns or seems to turn.
core (KOR) The center layer of a planet.
galaxies (GAH-lik-seez) Large groups of stars and the planets that circle them.
orbit (OR-bit) When one thing circles another.
satellite (SA-tel-yt) A heavenly body that revolves around a planet.
solar system (SOH-ler SIS-tem) A group of planets that circle a star. Our solar system has nine planets that circle the Sun.
space probe (SPAYS PROHB) A spacecraft that travels in space and is steered by scientists on the ground.
telescopes (TEL-uh-skohps) Tools used to make distant objects appear closer and larger.
temperature (TEM-pruh-cher) How hot or cold something is.
universe (YOO-nih-vers) Everything that is around us.

To learn more about Saturn, check out these Web sites:
http://spaceplace.jpl.nasa.gov/site_index.htm
The Web site below has information about the *Cassini* mission:
http:www.jpl.nasa.gov/cassini